Introduction

Welcome to this exciting puzzle book, crafted esp
who have dyslexia. Every puzzle in this book has
keeping in mind the unique strengths and challeı
activities not only offer fun and engagement but
their skills, boost their confidence, and develop a

Benefits of Each Puzzle:

1. Word Search:
• Pattern Recognition: puzzles help recognizing patterns which is crucial for reading and spelling.
• Working Memory: Remembering the list of words while searching for them enhances short-term memory.
• Visual Scanning Skills: Word searches help to improve the skill of scanning texts efficiently.

2. Rhyme Matching:
• Phonological Awareness: Understanding how words sound and rhyme boosts reading abilities.
• Building Vocabulary: Exposing kids to new words in a fun way enriches their vocabulary.
• Memory and Recall: Matching pairs demand attention and memory skills.

3. Rhyme Finishers:
• Creative Thinking: Kids are encouraged to think outside the box.
• Phonological Skills: Completing rhymes improves auditory skills and understanding of word sounds.
• Comprehension: Kids learn to anticipate and understand the context.

4. Spot the Difference Images:
• Improves Visual Perception: Helps in identifying subtle differences, enhancing visual discernment.
• Attention to Detail: Trains the eye and mind to notice minute details, vital for reading and writing.
• Concentration: Requires focused attention, which is great for building concentration span.

5. Reading Comprehension:
• Reading Stamina: Encourages kids to read longer texts without getting overwhelmed.
• Understanding: Questions help in ensuring the text is understood, promoting deeper comprehension.
• Critical Thinking: Kids learn to think about what they've read, developing analytical skills.

May this book provide countless hours of joy, small challenges, and growth for your young learner. Embrace the journey, celebrate the small wins, and remember, every puzzle solved is a step forward on the path of learning and confidence. Happy puzzling!

Word Puzzle: Animals

H	H	J	K	B	X	D	O	L	P	H	I	N	G
R	O	T	D	O	G	S	B	X	N	C	D	U	H
P	H	I	P	S	U	K	B	U	H	Z	D	W	W
X	D	G	I	I	M	O	N	K	E	Y	U	N	F
M	B	E	M	D	T	O	R	J	L	T	T	C	L
D	C	R	O	C	O	D	I	L	E	G	V	A	A
Y	A	Z	E	E	O	N	I	D	P	C	X	D	M
E	T	J	F	P	X	A	X	X	H	I	P	O	I
K	B	D	P	U	R	T	E	N	A	M	I	C	N
W	G	F	S	Z	S	L	I	O	N	C	M	G	G
I	X	D	U	S	H	F	U	F	T	A	Q	R	O
C	Y	W	S	C	Q	R	V	K	J	T	M	M	C
N	D	W	L	I	Q	K	F	T	D	W	Z	M	Z
D	U	R	P	H	O	R	S	E	D	V	N	M	R

Words:

Cat   Dog   Lion

Tiger   Horse   Monkey

Dolphin   Elephant   Flamingo

Crocodile

Rhyme Matching: Animals

Cat	Hare
Horse	Feel
Bear	Frog
Seal	Peer
Dog	Bat
Mouse	Pool
Goat	Boat
Crow	Box
Shark	Pale
Lion	Course
Mule	Mail
Fox	Iron
Duck	Truck
Bee	Lawn
Swan	Tree
Owl	Towel
Whale	Show
Deer	Park
Snail	Bull
Gull	House

Rhyme Finishers: Animals

In the jungle, the lion did **roar**.
Sleeping soundly, he began to _____.

The bird flew up into the **sky**.
Flapping wings, up so ____.

The snake slithered, sleek and **long**.
In the tall grass, he felt so _____.

The rabbit hopped along the **trail**.
With fluffy fur and a cotton ____.

The monkey swung from tree to **tree**.
Full of joy, wild and ____.

The mouse dashed, seeking a **treat**.
In a cozy corner, he found a _____.

Words:

Peep, High, Strong, Snore, Free, Seat, Tail

Spot 5 differences: Animals

Comprehension: Animals

Lucy visited the zoo. She saw a tall giraffe eating leaves. There was a playful monkey swinging from tree to tree. She also watched a big elephant splash water with its trunk. Lucy's favorite was the colorful parrot that said, "Hello!"

Questions:

Where did Lucy go?
Answer: _____

What was the giraffe doing?
Answer: _____

What animal was swinging from tree to tree?
Answer: _____

What did the elephant do with the water?
Answer: _____

Which animal said, "Hello!"?
Answer: _____

Word Search: Space

C	B	X	R	N	Y	V	E	N	U	S	E	A	V
I	U	X	D	H	O	N	O	V	Z	O	D	S	P
C	I	N	H	R	T	R	V	P	G	T	G	A	O
A	G	O	C	W	H	E	R	O	C	K	E	T	D
J	G	S	S	M	G	I	Q	W	M	J	L	E	M
T	S	T	A	R	H	P	T	G	W	U	Q	L	O
F	Q	J	M	A	D	J	J	A	K	P	W	L	J
E	A	R	T	H	P	H	Q	L	S	I	G	I	J
L	V	A	S	T	R	O	N	A	U	T	I	T	U
E	U	R	E	T	B	U	A	X	L	E	E	E	Y
X	S	U	N	W	L	Q	F	Y	C	R	V	R	W
Q	Q	Q	S	S	E	F	M	I	U	I	B	K	D
A	C	J	X	K	B	T	S	I	H	A	J	Q	S
J	M	O	O	N	O	N	S	J	N	R	D	F	U

Words:

Sun   Moon   Star
Earth   Venus   Rocket
Galaxy   Jupiter   Satellite
Astronaut

Rhyme Matching: Space

Moon	Mole
Stars	Far
Sun	Clips
Space	Hearth
Comet	Scuba
Planet	Bonnet
Mars	Taxi
Sky	Pocket
Astro	Bars
Rocket	Detour
Galaxy	Face
Nebula	Cars
Orbit	Metro
Star	Genius
Meteor	Granite
Blackhole	Fun
Saturn	Fly
Venus	Pattern
Earth	Spoon
Eclipse	Permit

Rhyme Finishers: Space

The stars in the sky shimmered **bright**.
Lighting up the deep dark _____.

The spaceship travelled at the speed of **light**.
Zooming forward, what a _____!

The planets orbited around the **sun**.
In the vast universe, they _____.

The comet left a trail so **fine**.
Through the stars, it did _____.

The astronaut floated, feeling no **weight**.
In space, he felt so _____.

The black hole was a mystery so **deep**.
Where time and space seem to _____.

Words:

Sleep, Sight, Night, Shine, Spun, Great

Spot 5 differences: Space

Comprehension: Space

Max looked up at the sky at night. He saw bright stars twinkling. A round moon glowed softly. He spotted a shooting star and made a wish. Max dreams of being an astronaut.

Questions:

When did Max look at the sky?
Answer: _____

What were the bright things Max saw in the sky?
Answer: _____

How did the moon look?
Answer: _____

What did Max do when he saw a shooting star?
Answer: _____

What is Max's dream job?
Answer: _____

Word Search: Nature

J	I	H	R	Y	I	Z	H	R	O	F	G	V	A	T	B
G	O	U	L	G	E	F	J	N	G	K	P	R	H	Q	Q
L	B	W	Y	M	L	Y	R	S	R	B	P	I	Q	P	Y
S	W	E	G	T	E	Y	L	X	S	U	E	V	O	I	D
G	Y	D	C	A	A	V	J	I	R	A	N	E	A	E	P
Z	L	J	Q	K	F	Y	A	C	T	B	G	R	A	S	S
Q	D	W	Z	B	A	F	J	M	R	H	M	A	P	F	L
M	O	K	T	M	I	E	B	P	T	W	K	I	R	H	B
H	V	X	N	Y	M	O	U	N	T	A	I	N	L	X	K
V	I	M	F	C	V	T	T	W	Z	M	D	B	G	P	W
B	R	J	O	R	C	O	T	D	I	T	L	O	T	Q	Q
N	R	I	R	O	T	R	E	E	A	P	R	W	Q	X	L
O	M	G	E	N	V	I	R	O	N	M	E	N	T	R	A
I	Y	I	S	T	I	N	F	L	O	W	E	R	R	C	F
I	L	Y	T	O	T	U	L	G	J	X	G	U	G	G	W
I	T	K	V	O	Q	V	Y	L	R	N	Q	B	I	R	Z

Words:

Tree	Leaf	Grass
River	Flower	Forest
Rainbow	Mountain	Butterfly
Environment		

Rhyme Matching: Nature

Tree	Cone
Flower	Power
Leaf	Crow
Rock	Dream
Stream	Sock
Sand	Kind
Stone	Cake
Cloud	Bun
Mountain	Pass
Hill	Fill
River	Ranch
Branch	Push
Bush	Pain
Grass	Liver
Wave	Fountain
Rain	Proud
Lake	Beef
Wind	Band
Snow	Cave
Sun	Bee

Rhyme Finishers: Nature

The tree's leaves wiggle and **sway**.
With the wind, they play all _____.

The river moves with a soft **sound**.
Over small stones, it winds _____.

Flowers pop up, red, blue, and **gold**.
Vibrant and beautiful, a wonder to _____.

Mountains stand, so high and **grand**.
Touching the clouds, like a magic _____.

Clouds look like shapes as they **float**.
Soft and fluffy, like a white _____.

Waterfalls drop with a big **splash**.
Making a sound, with a loud _____.

Words:

Around, Day, Land, Hold, Crash, Coat

Spot 5 differences: Nature

Comprehension: Nature

Lily went for a walk in the forest. She heard birds singing in the trees. A cool breeze rustled the leaves. By a stream, she saw a frog jump into the water. Everywhere, flowers bloomed in bright colors.

Questions:

Where did Lily go for a walk?
Answer: _____

What did she hear in the trees?
Answer: _____

What did the cool breeze do?
Answer: _____

What did the frog do by the stream?
Answer: _____

What bloomed in bright colors?
Answer: _____

Word Search: Food

A	U	B	I	E	N	F	X	C	M	M	S	V	X
S	M	W	G	R	X	I	F	O	X	W	G	C	N
F	R	B	K	F	E	B	A	B	P	J	X	H	D
T	O	R	A	N	G	E	R	L	C	W	L	E	I
W	M	E	C	J	O	R	K	U	H	S	R	E	Z
M	G	A	T	A	P	P	L	E	I	A	I	S	S
C	Y	D	R	K	R	K	T	B	C	N	C	E	W
C	H	O	C	O	L	A	T	E	K	D	E	I	N
G	O	J	O	B	N	H	T	R	E	W	Q	P	K
X	O	E	N	I	I	N	N	R	N	I	J	P	R
T	D	C	A	K	E	J	S	Y	E	C	J	I	P
P	D	T	I	R	G	N	N	S	Q	H	V	V	J
N	V	N	T	X	I	P	H	D	G	P	T	C	A
Z	K	I	X	Z	F	S	H	Z	U	D	F	Q	M

Words:

Rice	Cake	Apple
Bread	Orange	Cheese
Chicken	Sandwich	Chocolate
Blueberry		

Rhyme Matching: Food

Pie	Green
Bread	Loop
Rice	Nice
Meat	Moose
Soup	Please
Cheese	Beach
Bean	Cherry
Cake	Doodle
Fish	Sky
Fry	Lake
Juice	Demon
Steak	Cart
Grape	Sky
Lemon	Red
Berry	Beat
Tart	Ape
Noodle	Silk
Milk	Flake
Peach	Dish
Plum	Drum

Rhyme Finishers: Food

The apple was red, a treat to **eat**.
Crunchy and juicy, so very _____.

The pizza had cheese, stretchy and **nice**.
With toppings on it, like pepper and _____.

The cake was fluffy and oh so **sweet**.
With yummy icing, it's a treat to _____.

The soup warmed me, felt so **nice**.
On chilly days, it breaks the ____.

The ice cream was cold, a summer **treat**.
Melting fast in the sunny _____.

The sandwich was big, filled just **right**.
For my lunch, a tasty _____.

Words:

Eat, Spice, Sweet, Heat, Bite, Ice

Spot 5 differences: Foods

Comprehension: Food

Emma went to a fruit market with her mom. They picked juicy apples and ripe bananas. Emma also chose some red strawberries. At home, they made a tasty fruit salad. Emma loved the sweet taste.

Questions:

Where did Emma go with her mom?
Answer: _____

What fruits did they pick?
Answer: _____

What color were the strawberries Emma chose?
Answer: _____

What did they make at home?
Answer: _____

Did Emma enjoy the fruit salad?
Answer: _____

Word Search: Sports

G	Y	M	N	A	S	T	I	C	S	Q	B	I	K	E
W	L	I	W	E	F	E	B	N	K	C	A	D	D	T
H	I	W	C	R	U	N	N	T	A	L	S	L	X	Q
L	B	C	T	K	H	F	J	B	T	X	K	Y	F	M
D	A	W	K	M	H	T	M	V	E	A	E	D	V	N
D	T	E	N	N	I	S	H	O	B	L	T	Z	G	J
M	F	P	R	U	T	I	E	D	O	Z	B	L	W	S
X	N	Z	W	B	H	F	V	O	A	D	A	D	A	W
U	H	W	R	A	I	W	B	Q	R	U	L	I	N	I
Y	W	H	Q	L	M	A	R	C	D	C	L	G	F	M
L	P	M	F	L	H	C	N	M	W	J	U	Y	X	J
R	B	W	T	K	W	G	P	P	Y	P	Z	P	P	C
B	A	S	E	B	A	L	L	T	P	X	O	H	B	I
O	L	N	V	I	I	C	C	Z	A	H	D	T	O	J
W	H	P	R	L	A	S	J	H	R	Z	G	O	G	T

Words:

Bat	Run	Ball
Swim	Bike	Tennis
Baseball	Basketball	Gymnastics
Skateboard		

Rhyme Matching: Sports

Ball	Stick
Bat	Dating
Track	Hall
Race	Fort
Field	Rattle
Score	Sun
Swim	Turtle
Goal	Limb
Kick	Door
Dive	Shield
Tackle	Ditch
Run	Hole
Jump	Hive
Hit	Base
Pitch	Fit
Sport	Back
Skating	Bow
Hurdle	Lump
Dart	Hat
Row	Cart

Rhyme Finishers: Sports

The soccer ball rolled on the **grass**.
Into the goal, it went so ____.

The bat hit the ball with a loud **sound**.
Flying high, then it hit the _____.

The swimmer jumped into the **pool**.
Swimming fast, she was really ____.

The runner dashed, swift on his **feet**.
Hoping in the race, he wouldn't be ____.

The skater glided, smooth on the **ice**.
Twirling and spinning, it looked so _____.

The gymnast jumped, high and **neat**.
Landing on her two ____.

Words:

Feet, Fast, Cool, Ground, Nice, Beat

Spot 5 differences: Sports

Comprehension: Sports

Jake loves sports day at school. He ran in a race and came second. He kicked a football into the goal. Later, he tried jumping over a high bar. At the end of the day, Jake won a blue ribbon.

Questions:

Who loves sports day?
Answer: _____

How did Jake do in the race?
Answer: _____

What did he do with a football?
Answer: _____

What did Jake try jumping over?
Answer: _____

What did Jake win at the end of the day?
Answer: _____

Word Search: Transport

U	Y	T	B	M	O	D	W	Q	N	Q	P	F	F	V
J	I	R	N	O	N	J	Z	Y	A	I	B	U	S	L
A	F	A	Y	T	J	G	C	P	T	K	O	U	A	T
D	K	I	O	O	Z	D	S	C	O	O	T	E	R	I
P	H	N	S	R	W	W	H	O	K	J	D	Y	O	Z
S	X	G	K	C	A	R	S	U	B	W	A	Y	L	L
O	D	H	Q	Y	B	Y	R	M	C	G	P	Z	H	Q
H	E	L	I	C	O	P	T	E	R	W	T	Q	U	J
J	G	H	X	L	A	S	O	C	W	D	Z	V	N	D
Q	B	I	K	E	T	K	Q	P	P	M	Q	K	X	G
P	T	C	M	W	O	H	R	T	L	X	G	Q	X	W
R	C	Y	V	C	B	D	R	Q	A	G	Q	S	E	P
D	Y	U	H	M	Z	G	F	M	N	X	R	H	A	C
Q	Y	D	V	G	S	H	J	Y	E	S	J	J	B	Z
L	P	S	Z	U	Z	L	R	D	E	V	Q	T	H	Q

Words:

Car	Bus	Bike
Boat	Train	Plane
Subway	Scooter	Helicopter
Motorcycle		

Rhyme Matching: Transport

Car	Peep
Train	Retro
Boat	Tricycle
Plane	Sky
Truck	Computer
Ride	Flip
Fly	Taught
Bike	Maxi
Van	Crane
Bus	Plus
Ship	Hike
Road	Dive
Rail	Star
Drive	Snail
Taxi	Slide
Metro	Toad
Yacht	Rain
Cycle	Coat
Scooter	Duck
Jeep	Pan

Rhyme Finishers: Transport

The plane soared up, touching the **sky**.
Clouds all around, flying so _____.

The bike pedaled forward, feeling the **breeze**.
Riding down pathways with so much _____.

The plane soared up, way up **high**.
Above the hills, where eagles _____.

The bicycle rode down the **lane**.
Quick and fun, even in the _____.

The boat moved gently on the blue **sea**.
Sailing along, as free as can _____.

The bus stopped, here and **there**.
Picking up people, from _____.

Words:

Fly, Ease, High, Be, Everywhere, Rain

Spot 5 differences: Transport

Comprehension: Transport

Sammy loves watching vehicles. He sees big buses drive by. He waves at the colorful bicycles on the path. Every day, he hears a train whistle from afar. His favorite is the blue airplane that flies high in the sky.

Questions:

What does Sammy love to watch?
Answer: _____

What does he see drive by?
Answer: _____

What does Sammy wave at on the path?
Answer: _____

What does he hear every day?
Answer: _____

What is Sammy's favorite vehicle?
Answer: _____

Word Search: School

Q	R	Q	A	O	P	J	S	C	R	H	O	H	J	R
O	W	T	O	E	L	F	O	A	L	B	P	C	T	X
R	X	J	O	D	V	H	A	B	E	K	I	V	G	M
C	T	F	Z	G	W	Z	U	W	P	U	P	T	L	A
G	P	W	S	Z	D	T	B	J	I	O	A	T	R	R
K	K	Z	G	X	V	E	L	D	O	J	P	Q	X	L
G	Q	M	V	X	Z	A	A	T	J	T	E	M	E	Q
C	W	Z	R	P	C	C	C	H	A	I	R	F	D	L
P	L	O	A	D	N	H	K	D	T	K	A	A	R	M
E	L	E	E	E	O	E	B	O	O	K	W	M	U	V
N	C	L	A	S	S	R	O	O	M	U	L	E	B	Y
C	B	N	G	K	C	C	A	R	I	X	H	P	W	O
I	U	P	L	A	Y	G	R	O	U	N	D	E	K	U
L	C	R	A	Y	M	Z	D	Z	X	M	Q	N	S	K
I	H	H	E	Z	J	D	T	P	C	U	P	W	P	N

Words:

Pen	Book	Desk
Chair	Paper	Pencil
Teacher	Classroom	Playground
Blackboard		

Rhyme Matching: School

Book	Fern
Pen	Path
Chair	Blade
Desk	Hen
Note	Chaser
Test	Preacher
Class	Seed
Math	Whizz
Learn	Look
Read	Best
Pencil	Well
Quiz	Glass
Teacher	Prudent
Bell	Nest
Board	Ford
Grade	Bear
Paper	Boat
Student	Cresson
Eraser	Taper
Lesson	Stencil

Rhyme Finishers: School

The bell rang, Books and bags, tools so **neat**.
Ready for lessons, kids took their _____.

Notebooks open, pens ready to **glide**.
Learning and fun, side by ____.

The teacher wrote words, easy to **see**.
Learning is fun for you and ____.

On the playground, we play and **run**.
Laughter and games, everyone having ____.

Line up straight, wait for the **bell**.
Another day done, stories to ____.

Outside it stands, the bus so **tall**.
Ready to drive, kids big and ____.

Words:

Side, Me, Seat, Tell, Fun, Small

Spot 5 differences: School

Comprehension: School

Anna goes to Green Valley School. Every day, she carries a yellow backpack. Inside, there are books, pencils, and a lunchbox. Her favorite class is art where she paints pictures. Anna's best friend, Mia, sits next to her.

Questions:

Where does Anna go every day?
Answer: _____

What color is her backpack?
Answer: _____

What does Anna carry inside her backpack?
Answer: _____

What is Anna's favorite class?
Answer: _____

Who sits next to Anna?
Answer: _____

Word Search: Weather

C	I	B	S	B	Q	H	H	S	F	T	D	Q	G	M	X
W	V	S	S	Q	I	Q	T	M	Z	H	C	D	L	O	M
Q	L	T	E	M	P	E	R	A	T	U	R	E	I	O	W
V	X	O	F	G	M	D	R	A	I	N	Z	L	G	V	B
A	O	R	M	N	B	C	L	O	U	D	T	R	H	F	H
U	T	M	C	M	A	A	Z	Q	L	E	Y	P	T	V	J
K	J	A	L	Y	Y	M	T	P	P	R	C	Y	N	E	O
B	L	I	Z	Z	A	R	D	B	K	E	K	N	I	K	C
F	E	F	A	A	T	P	E	I	L	S	T	N	N	H	H
U	M	C	A	L	N	Z	K	M	C	O	O	F	G	M	R
T	G	O	X	Y	D	A	Z	E	C	D	Y	X	P	N	E
E	E	N	J	V	S	X	C	W	H	F	W	I	N	D	R
U	V	E	P	P	T	K	Z	X	G	E	K	D	D	U	U
M	K	H	W	O	U	U	C	H	D	E	S	U	N	U	O
K	R	E	U	K	M	W	U	G	T	S	N	O	W	E	P
Q	B	M	W	Y	X	O	B	V	X	X	Y	A	Q	F	N

Words:

Sun	Rain	Wind
Snow	Cloud	Storm
Thunder	Blizzard	Lightning
Temperature		

Rhyme Matching: Weather

Rain	Pale
Snow	Warm
Cloud	Train
Wind	Lizard
Storm	Fizzle
Fog	Beat
Mist	Frightening
Sun	Shoe
Thunder	Trees
Lightning	Avocado
Drizzle	Hill
Chill	Pinned
Breeze	Fun
Hail	Flower
Frost	Under
Tornado	Lost
Heat	Dog
Shower	List
Blizzard	Loud
Dew	Bow

Rhyme Finishers: Weather

The sun shone bright, warming the **ground**.
Casting shadows all _____.

Days are short, when winter is **here**.
Nights are long, and the cold is _____.

Puddles form, when it rains all **day**.
Boots splash in, as kids ____.

Mist in the morning, fades with **light**.
As the sun rises, everything's _____.

The thunder roared, lighting up the **sky**.
With every flash, clouds fly ___.

The fog enveloped, everything in **sight**.
Mysterious and quiet, like the night.

Words:

By, Play, Near, Bright, Night, Around

Spot 5 differences: Weather

Comprehension: Weather

Oscar looked out the window. Today, the sun was shining brightly. A gentle wind made the trees sway. Puffy clouds floated in the blue sky. Oscar hoped it would rain tomorrow for the plants.

Questions:

What did Oscar do?
Answer: _____

How was the sun today?
Answer: _____

What did the wind do to the trees?
Answer: _____

What floated in the blue sky?
Answer: _____

What did Oscar hope for tomorrow?
Answer: _____

Word Search: Holidays

C	Z	N	O	N	P	A	V	L	U	K	S	F	H	Y
P	U	M	P	K	I	N	T	M	V	E	Z	N	Q	W
Y	L	B	X	R	L	J	Y	Y	Z	O	O	R	H	S
L	L	U	P	B	X	X	S	B	Y	Y	Y	B	X	T
Q	E	Q	P	P	F	B	O	U	Z	R	P	J	E	K
D	C	A	F	U	S	T	A	R	N	E	T	T	M	K
T	K	E	O	M	M	H	N	R	C	I	R	S	P	I
D	Q	F	U	O	U	X	J	W	H	N	E	P	H	J
V	E	R	K	P	C	D	H	L	O	D	E	V	G	S
C	G	W	P	C	P	B	B	I	C	E	V	Z	V	P
A	G	A	I	A	P	E	T	J	O	E	F	E	P	Z
K	C	Z	N	N	Y	V	D	B	L	R	W	K	J	S
N	M	L	N	D	E	C	O	R	A	T	I	O	N	J
G	S	K	B	Y	G	H	O	S	T	U	R	K	E	Y
E	A	W	A	R	M	P	H	V	E	Q	H	J	Q	I

Words:

Egg
Candy
Pumpkin
Decoration

Star
Ghost
Reindeer

Tree
Turkey
Chocolate

Rhyme Matching: Holidays

Gift	Dandy
Tree	Shelf
Treat	Least
Feast	Yard
Card	Atlanta
Snow	Well
Bell	Dove
Santa	Plume
Elf	Dunkin
Candy	Lift
Easter	Made
Pumpkin	Ditch
Ghost	Feaster
Witch	Bow
Turkey	Toast
Parade	Sweet
Firework	Bee
Love	Quirk
Costume	Murky
Bunny	Sunny

Rhyme Finishers: Holidays

The presents sat under the Christmas **tree**.
Beatufully wrapped for all to ____.

New Year's Eve, counting down **loud**.
Welcoming the new year, with a ____.

Green all over, St. Patrick's **fun**.
 Looking for clovers, one by ____.

Birthday cakes, candles to **blow**.
Make a wish, and let it ____.

Plates are full, with festive **cheer**.
Celebrating the end of the ____.

Costumes on, a spooky **sight**.
Trick or treat, its Halloween ____.

Words:
Grow, See, One, Crowd, Night, Year

Spot 5 differences: Holidays

Comprehension: Holidays

Sophie loves the holidays. In December, she sees twinkling lights on houses. She helps bake gingerbread cookies. There's a tall pine tree in her living room with shiny ornaments. On Christmas morning, she finds presents wrapped in colorful paper.

Questions:

Who loves the holidays?
Answer: _____

What does Sophie see on houses in December?
Answer: _____

What kind of cookies does she help bake?
Answer: _____

Where is the tall pine tree?
Answer: _____

What does she find on Christmas morning?
Answer: _____

Word Search: Fantasy

K	K	W	D	Q	W	X	O	V	A	J	U	L	S
V	F	F	B	N	O	M	S	M	H	G	X	Y	N
D	Q	R	C	C	A	S	T	L	E	I	F	W	A
C	K	S	Y	F	C	S	Y	P	F	M	L	A	D
S	H	J	J	L	V	B	D	R	A	G	O	N	V
Q	Q	E	Y	C	S	U	W	I	Z	W	A	D	E
N	L	L	Y	A	X	Q	W	N	T	Q	R	L	N
G	X	F	J	M	A	G	I	C	C	M	R	W	T
O	Y	A	F	W	W	X	J	E	T	Y	X	K	U
S	K	I	N	G	C	R	S	S	O	Q	N	B	R
S	O	R	C	E	R	E	S	S	U	P	T	U	E
O	R	Y	F	O	A	Y	S	K	K	Y	W	L	X
I	P	V	A	G	R	N	W	O	B	S	S	C	Q
P	I	P	M	W	I	M	V	O	G	U	M	G	B

Words:

Elf	Magic	Princess
Wand	Castle	Adventure
King	Dragon	Sorceress
Fairy		

Rhyme Matching: Fantasy

Knight	Berry
Dragon	Green
Spell	Foam
Fairy	Defiant
Castle	Pond
Giant	Shelf
Wand	Tassel
Mermaid	Ocean
Unicorn	Scarf
Gnome	Tragic
Potion	Torn
King	Cruel
Queen	Wagon
Troll	Light
Dwarf	Clown
Elf	Ring
Ogre	Bell
Jewel	Parade
Magic	Poker
Crown	Roll

Rhyme Finishers: Fantasy

The dragon flew over the castle's **peak**.
Breathing fire, with a roar and a ____.

The fairy sprinkled magic dust in the **air**.
Sparkling and shimmering, _____.

Unicorns run, with horns so **rare**.
Rainbows shine, in the ___.

The mermaid sang songs deep in the **sea**.
With shimmering scales, as free as can ____.

The wizard waved his wand, spells to **cast**.
Changing the present, future, and ____.

A magic carpet, takes to the **sky**.
Over lands and seas, up so ____.

Words:

Everywhere, Air, Shriek, Past, Be, High

Spot 5 differences: Fantasy

Comprehension: Fantasy

In a land of magic, Ellie found a sparkling wand. With a swish, she turned a rock into gold. Flying dragons soared above, and talking animals chatted by her side. Ellie's adventure in the enchanted forest was just beginning.

Questions:

Where did Ellie find a sparkling wand?
Answer: _____

What did she turn into gold?
Answer: _____

What flew above in the sky?
Answer: _____

Who could talk in this magical land?
Answer: _____

Where was Ellie's adventure taking place?
Answer: _____

Word Search: Dinosaur

R	O	A	R	I	Z	W	H	K	G	V
S	G	L	N	F	R	U	W	T	A	V
T	R	F	O	S	S	I	L	C	N	L
R	A	B	Z	T	R	V	O	B	S	S
I	P	X	U	E	A	F	T	I	T	O
C	T	A	K	G	C	W	F	X	X	N
E	O	Y	B	O	D	I	N	O	E	Z
R	R	Q	O	K	C	L	A	W	S	B
H	S	B	N	X	V	R	X	B	J	P
K	T	R	E	X	C	T	A	I	L	Q
O	A	L	L	V	B	P	V	K	I	K

Words:

Rex	Dino	Bone
Fossil	Raptor	Tricer
Tail	Claw	Stego
Roar		

Rhyme Matching: Dinosaurs

Rex	Sing
Bone	Song
Raptor	Dusk
Tail	Test
Claw	Straw
Stego	Put
Roar	Beneath
Spike	Pale
Foot	Sky
Scale	Stunt
Dino	Actor
Beak	Child
Wing	Snail
Fly	Peek
Tusk	Vino
Hunt	Hike
Teeth	Cone
Wild	Flex
Long	Door
Nest	Lego

Rhyme Finishers: Dinosaurs

Long ago, dinosaurs walked and **played**.
T-Rex roared, making others _____.

Triceratops had horns, looking **neat**.
It loved to eat plants, its favorite _____.

Pterosaurs flew up, way up **high**.
With big wings, they touched the _____.

Hidden dino eggs, safe and **sound**.
Inside them, new life is _____.

Pterosaurs soared, wide wings in **glide**.
Over land and rivers, they took _____.

Triceratops had horns, one, two, **three**,
Munching on plants, as happy as can ___.

Words:

Treat, Sky, Afraid, Found, Be, Pride

Spot 5 differences: Dinosaur

Comprehension: Dinosaurs

Danny loves reading about dinosaurs. The T-Rex had sharp teeth and was very big. The gentle Brachiosaurus ate plants and had a long neck. He also learned about the speedy Velociraptor. Danny's room has dinosaur posters everywhere.

Questions:

What does Danny love to read about?
Answer: _____

What did the T-Rex have?
Answer: _____

What did the Brachiosaurus eat?
Answer: _____

Which dinosaur was speedy?
Answer: _____

What does Danny have in his room?
Answer: _____

Word Search: Under the sea

N	M	Z	U	W	I	N	W	M	I
Q	R	Z	E	I	O	L	R	R	M
H	K	S	E	H	D	W	A	V	E
S	B	R	F	I	S	H	Z	K	W
P	N	S	C	O	R	A	L	K	Q
H	D	I	R	R	S	L	N	P	Y
I	I	U	A	S	H	E	L	L	A
W	V	Q	B	A	A	T	L	G	F
Q	E	N	N	N	R	I	J	W	H
R	E	E	F	D	K	M	F	J	Q

Words:

Fish
Whale
Crab
Shark

Sand
Shell
Dive

Coral
Wave
Reef

Rhyme Matching: Under the Sea

Fish	Thief
Sand	Moral
Coral	Hive
Whale	Hide
Shell	Lid
Wave	Park
Crab	Far
Dive	Pale
Reef	Coat
Shark	Dish
Squid	Brave
Swim	Play
Boat	Bell
Fin	Limb
Tide	Tin
Star	Goat
Deep	Hand
Float	Shoe
Blue	Slab
Ray	Note

Rhyme Finishers: Under the Sea

Fish swim below, where the seaweeds **sway**,
Coral and plants light up their ____.

A mermaid's song, clear and **bright**,
Filling the sea, day and ____.

Dolphins jump, up so **high**,
Almost touching, the blue ____.

Starfish rest, and seahorses **play**,
In the sea, they dance and ____.

Starfish lay, on the sea's **ground**,
While seahorses, swirl ____.

Sharks zoom by, swift and **sleek**,
In the big sea, they play hide and ____.

Words:

Around, Seek, Night, Way, Sway, Sky

Spot 5 differences: Under the Sea

Comprehension: Under the Sea

Molly went snorkeling in the clear blue ocean. She saw colorful fish darting around. A graceful jellyfish floated by. Nearby, a playful dolphin leaped out of the water. Under the sea, Molly felt like she was in another world.

Questions:

What did Molly do in the ocean?
Answer: _____

What kind of fish did she see?
Answer: _____

What floated by Molly?
Answer: _____

Which sea creature leaped out of the water?
Answer: _____

How did Molly feel under the sea?
Answer: _____

Word Search: Superheroes

Q	N	A	B	R	O	Z	Z	A	L
T	H	E	R	O	M	S	N	G	K
H	I	N	A	N	N	A	U	O	A
A	W	A	V	X	L	V	L	U	N
S	P	E	E	D	L	E	P	P	V
F	I	G	H	T	V	R	R	O	T
L	C	Y	S	E	P	M	B	W	E
Y	M	S	T	Y	C	A	P	E	A
T	O	Q	H	Y	S	S	U	R	M
S	G	H	T	O	O	K	B	L	Z

Words:

Cape
Hero
Save
Fight

Mask
Power
Team

Fly
Speed
Brave

Rhyme Matching: Superheroes

Cape	Ebb
Mask	Cave
Fly	Ring
Hero	Park
Power	Reed
Speed	Wire
Save	Stick
Team	Light
Brave	Tower
Fight	Brave
Wing	Dream
Quest	Salt
Night	Best
Flash	Sky
Web	Feel
Fire	Flask
Quick	Tape
Steel	Zero
Dark	Dash
Vault	Right

Rhyme Finishers: Superheroes

With a cape and mask, into the **sky**,
Our superhero began to ____.

A city in peril, a signal in **air**,
Our hero arrives, with time to ____.

A super suit, shiny and **new**,
Villains beware, they're after ____.

Saving the day, without a **rest**,
Every child knows they're the ____.

Faster than light, swift as a **breeze**,
Jumping over skyscrapers with ____.

Invisible planes, shields that **throw**,
Superheroes are always on the ____.

Words:

Ease, Fly, Spare, Best, Go, You

Spot 5 differences: Superheroes

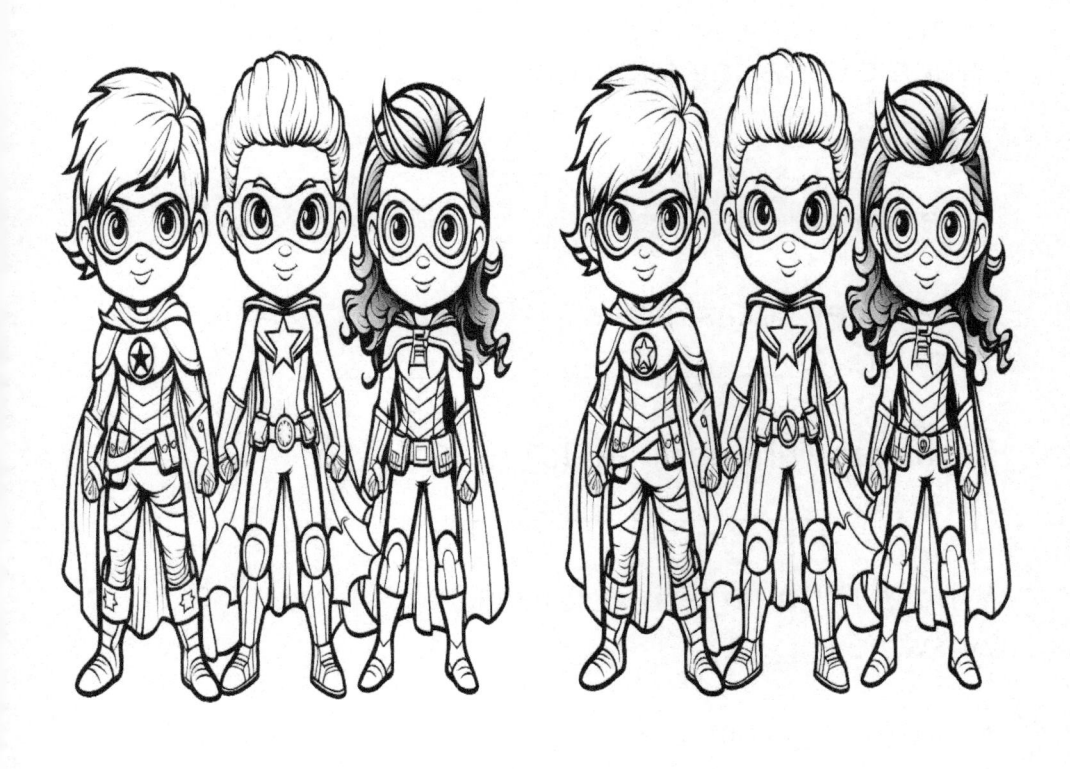

Comprehension: Superheroes

Lucas has a superhero comic book. The hero, ThunderMan, can fly and shoot lightning bolts. His sidekick, FlashGirl, runs faster than the wind. Together, they save the city from villains. Lucas dreams of having superpowers too.

Questions:

What does Lucas have?
Answer: _____

What can ThunderMan do?
Answer: _____

Who is ThunderMan's sidekick?
Answer: _____

How fast does FlashGirl run?
Answer: _____

What does Lucas dream of?
Answer: _____

Word Search: Farm Life

N	Y	N	E	S	B	B	Q	C	Q	G	V
P	L	D	L	D	V	C	A	U	Z	F	R
M	P	Z	X	G	O	A	T	H	P	J	C
U	Q	A	R	H	H	F	N	X	W	J	O
T	R	A	C	T	O	R	B	A	R	N	R
B	O	D	Y	S	R	O	C	O	G	E	N
T	D	U	B	H	S	C	X	O	A	I	E
I	N	C	O	E	E	O	D	W	L	Y	R
M	P	K	H	E	D	W	E	J	I	E	V
U	X	O	T	P	I	G	U	B	O	K	J
U	R	J	S	V	E	X	Z	P	R	R	G
Y	X	F	A	R	M	B	D	H	A	Y	W

Words:

Cow	Pig	Farm
Sheep	Barn	Tractor
Corn	Duck	Horse
Goat		

Rhyme Matching: Farm Life

Cow	Rain
Hay	Arm
Pig	Stick
Barn	Truck
Duck	Pen
Corn	Yarn
Goat	Leap
Horse	Plow
Sheep	Day
Chick	Laugh
Mud	Bowl
Farm	Horn
Hen	Seed
Stable	Table
Calf	Course
Field	Bud
Foal	Actor
Tractor	Boat
Grain	Twig
Feed	Shield

Rhyme Finishers: Farm Life

Cows in the pasture, munching on **hay**,
The farmer works hard, all ____.

Pigs in the mud, happy and **free**,
Chickens lay eggs, for you and ____.

The tractor hums, fields to **plow**,
Farm life is peaceful, here and ____.

Fresh veggies grow, in the farmer's **plot**,
Juicy tomatoes, and carrots a _____.

Kittens in the barn, playing in **hay**,
On the farm, it's a sunny ____.

Pigs in the mud, having some **fun**,
Cooling off, under the ____.

Words:

Now, Me, Day, Lot, Sun, Day

Spot 5 differences: Farm Life

Comprehension: Farm Life

Tom lives on a farm. He has a big red barn. Every morning, he feeds the chickens and collects their eggs. In the afternoon, he plays with his dog, Spot, near the pond. Tom loves farm life.

Questions:

Where does Tom live?
Answer: _____

What color is the barn?
Answer: _____

What does Tom do every morning?
Answer: _____

Who does Tom play with in the afternoon?
Answer: _____

Does Tom like living on the farm?
Answer: _____

Word Search: Pirates

O	B	B	P	L	A	N	K	J	T
R	O	U	W	P	B	O	Q	P	H
G	O	F	C	S	H	I	P	M	O
T	T	L	H	U	C	E	E	K	O
F	Y	Q	E	O	R	J	W	C	K
U	G	U	S	F	E	B	T	H	M
H	S	V	T	L	W	K	O	B	A
L	E	W	S	A	I	L	V	A	P
R	R	G	C	G	O	L	D	V	Z
G	J	I	M	N	J	F	R	I	D

Words:

Ship
Chest
Crew
Sail

Gold
Hook
Plank

Map
Flag
Booty

Rhyme Matching: Pirates

Ship	Pile
Gold	Look
Chest	Fly
Booty	Cap
Plank	Full
Flag	Bold
Map	Peck
Pirate	Lanker
Sword	Potion
Hook	Private
Sail	Carrot
Anchor	Boot
Isle	Pale
Loot	Board
Deck	Crate
Mate	Rank
Parrot	Best
Skull	Bag
Ocean	Cutie
Spy	Slip

Rhyme Finishers: Pirates

With a map and compass in his **hand**,
The pirate searched for treasure _____.

Sailing the seas, so wild and **blue**,
The pirate crew sang songs, old and ____.

The Jolly Roger, high it **flies**,
Pirates under the open _____.

The island ahead, where treasures **hide**,
With a pirate song, they sail with _____.

Beware the ghost ship, sailing at **night**,
With ghostly pirates, ready to _____.

The compass points, to adventure and **fun**,
Under the Caribbean ____.

Words:

Skies, New, Pride, Land, Sun, Fight

Spot 5 differences: Pirates

Comprehension: Pirates

Jack found an old map in his attic. It showed an island with an "X" mark. Dreaming of buried treasure, he wore a pirate hat and pretended his bed was a ship. With his toy parrot, Jack set sail on an imaginary adventure.

Questions:

What did Jack find in his attic?
Answer: _____

What was marked on the island?
Answer: _____

What did he wear on his head?
Answer: _____

What did he pretend his bed was?
Answer: _____

Who accompanied Jack on his adventure?
Answer: _____

Word Search: Construction Site

A	V	D	K	C	V	M	F	C	W
C	B	U	I	L	D	O	D	G	H
C	R	A	N	E	I	N	I	F	R
A	I	Z	A	H	G	V	B	Q	R
E	C	S	I	T	E	N	E	G	G
S	K	Y	L	H	I	G	A	H	R
Q	W	G	T	O	O	L	M	Q	O
U	G	T	T	R	U	C	K	A	A
G	Q	U	T	L	T	L	I	O	D
N	F	H	U	J	P	L	C	C	G

Words:

Dig	Crane	Brick
Nail	Road	Build
Tool	Truck	Beam
Site		

Rhyme Matching: Construction Site

Brick	Shirt
Crane	Stick
Dig	Jolt
Tool	Rent
Cement	Rig
Beam	Drift
Nail	Lurk
Hammer	School
Truck	Wipe
Build	Luck
Dirt	Frill
Block	Clock
Drill	Lane
Weld	Tail
Bolt	Blue
Lift	Load
Pipe	Held
Road	Filled
Work	Dream
Crew	Stammer

Rhyme Finishers: Construction Site

From dawn till dusk, they work with **pride**,
At the construction site, side by ____.

The foreman checks, plans in **hand**,
Ensuring everything's built as ____.

Concrete pours, setting the **base**,
Every brick finds its ____.

Cranes lift high, touching the **sky**,
Building dreams, way up ____.

Blueprint plans, held so **tight**,
Making sure every bolt is ____.

The foundation strong, walls so **neat**,
Soon a new building, stands on the ____.

Words:

Planned, Right, Place, High, Street, Side

Spot 5 differences: Construction Site

Comprehension: Construction Site

Liam watched the construction site from his window. Big trucks carried loads of dirt. A tall crane lifted heavy beams. Workers in helmets drilled and hammered. Every day, a new building grew taller. Liam imagined being a builder too.

Questions:

What did Liam watch from his window?
Answer: _____

What did the big trucks carry?
Answer: _____

What did the tall crane lift?
Answer: _____

What did the workers wear on their heads?
Answer: _____

What did Liam imagine himself as?
Answer: _____

Word Search: Robots

K	M	M	H	J	P	M	N	B	D	X
A	B	O	L	T	B	O	T	G	G	A
M	U	T	F	T	W	I	R	E	P	I
F	C	O	D	E	A	R	W	A	A	G
S	C	R	E	E	N	Z	Q	R	M	F
Y	H	K	P	U	C	Y	F	U	Y	B
N	I	Z	Q	O	R	O	J	T	C	D
M	P	P	N	N	L	A	S	E	R	W
S	Y	E	S	D	G	H	C	C	D	P
O	S	T	Y	G	N	Q	A	H	A	L
S	X	Q	A	Z	Z	W	E	Z	Y	M

Words:

Bot	Gear	Wire
Chip	Motor	Screen
Code	Tech	Bolt
Laser		

Rhyme Matching: Robots & Future

Wire	Dash
Chip	Thin
Bot	Green
Code	Jolt
Screen	Dream
Gear	Cot
Tech	Slip
Beam	Fight
Light	Lid
Port	Hug
Byte	Hive
Disk	Sight
Plug	Fire
Drive	Risk
Bolt	Load
Spin	Fort
Flash	False
Pulse	Ear
Track	Deck
Grid	Back

Rhyme Finishers: Robots & Future

In a world of steel, circuits and **gear**,
Robots move, without any ____.

With flashing lights and voices so **clear**,
They perform tasks, year after ____.

Flying cars, and cities so **grand**,
The future is here, just as _____.

Machines and gadgets, sleek and **new**,
The future's bright, for me and ____.

Artificial intelligence, thinking so **fast**,
Memories of the present, and of the _____.

Electric cars, zooming **around**,
Silent and smooth, without a _____.

Words:

Year, Past, Planned, Sound, You, Fear

Spot 5 differences: Robots

Comprehension: Robots

Emily has a toy robot named Rizo. Rizo can dance, light up, and even speak simple words. Its eyes are bright blue LEDs. When Emily presses a button, Rizo spins around. Every night, she places Rizo on its charging dock.

Questions:

What does Emily have named Rizo?
Answer: _____

What can Rizo do?
Answer: _____

What color are Rizo's eyes?
Answer: _____

What happens when Emily presses a button?
Answer: _____

Where does Emily place Rizo every night?
Answer: _____

Word Search: Jungle

K	T	H	W	E	S	L	P	B	X	B
D	R	U	O	V	F	S	P	C	D	I
U	R	Z	R	J	U	N	G	L	E	R
O	J	I	F	T	C	A	T	C	F	D
O	Q	Z	K	L	W	K	F	V	O	L
C	W	Q	R	I	V	E	R	G	U	K
C	T	Z	H	O	E	U	O	Q	S	G
N	C	V	I	N	E	A	G	J	L	E
L	L	Z	L	E	A	F	G	Q	L	B
Q	C	A	N	P	P	X	Q	J	V	B
O	Q	U	S	S	E	J	Z	L	P	Z

Words:

Cat	Lion	Vine
Snake	Frog	Bird
Leaf	Ape	River
Jungle		

Rhyme Matching: Jungle Adventure

Lion	Log
Vine	Word
Monkey	Striker
Tiger	Nut
Leaf	Bungle
Snake	Donkey
Jungle	Straw
Bird	Quiver
Tribe	Vibe
Drum	Gum
Plant	Child
River	Chant
Path	Pine
Rain	Oar
Hut	Ryan
Frog	Ring
Swing	Bath
Roar	Cane
Claw	Lake
Wild	Beef

Rhyme Finishers: Jungle Adventure

Monkeys swing, from vine to **tree**,
Jungle life, wild and ____.

The river flows, with secrets **deep**,
Where crocodiles and hippos _____.

Exotic birds, with colors so **fine**,
Fly above the thick tree _____.

A hidden temple, ancient and **old**,
Jungle tales, waiting to be _____.

The roar of the lion, king of the **land**,
Over his pride, he takes a _____.

Vines swing freely, from tree to **tree**,
Monkeys in motion, as wild as can ____.

Words:

Line, Free, Told, Stand, Be, Sleep

Spot 5 differences: Jungle

Comprehension: Jungle

Oliver went on a jungle safari with his family. Tall trees stretched up to the sky. He heard the roar of a lion and the chatter of monkeys. Colorful birds flew overhead. In the distance, an elephant splashed in a river. The jungle was full of life.

Questions:

Where did Oliver go with his family?
Answer: _____

What stretched up to the sky?
Answer: _____

What sounds did Oliver hear in the jungle?
Answer: _____

Which animals splashed in the river?
Answer: _____

Was the jungle empty or full of life?
Answer: _____

Word Search: Music

H	X	R	D	R	M	T	N	Q	R	S
S	A	K	Y	N	E	J	O	Z	N	B
C	Y	C	S	F	L	U	T	E	R	A
A	H	H	R	I	O	B	E	A	T	N
O	B	O	I	P	D	I	H	R	U	D
C	A	R	C	D	Y	Q	S	E	N	F
V	S	D	F	P	I	E	I	X	E	D
S	B	Y	S	O	N	G	D	E	H	R
S	Z	H	N	S	A	N	O	E	F	U
L	M	S	E	W	T	T	P	G	Y	M
B	U	B	A	S	S	U	K	W	D	D

Words:

Note
Song
Band
Melody

Drum
Flute
Tune

Beat
Chord
Bass

Rhyme Matching: Music

Note	Seat
Drum	Long
Beat	Ditch
Song	Has
Play	Cap
Flute	Boot
String	Ring
Chord	Board
Band	Day
Tune	Moon
Sing	Remedy
Bass	Coat
Loud	Hop
Pitch	Sock
Rock	Grass
Jazz	Ring
Blues	Cloud
Pop	Plum
Rap	Shoes
Melody	Sand

Rhyme Finishers: Music

The piano plays, notes so **clear**,
Melodies that we hold ____.

Dance to the beat, feel the **groove**,
With every song, our spirits ____.

A rock band plays, loud and **clear**,
The crowd goes wild, full of ____.

The opera singer, hits a high **note**,
Music floats, like a ____.

Harmonicas, trumpets, and **bass**,
Music fills every ____.

Pianist's fingers, dance on **keys**,
Music that flows, like a gentle ____.

Words:
Cheer, Move, Boat, Dear, Breeze, Space

Spot 5 differences: Music

Comprehension: Music

Ava loves to listen to music. Every morning, she plays her small ukulele. She taps her feet to the rhythm of a drum. At school, she sings songs with her friends. Her favorite instrument is the shiny piano at home. Music makes Ava's world brighter.

Questions:

What does Ava love to do?
Answer: _____

What instrument does she play every morning?
Answer: _____

What does she do to the rhythm of a drum?
Answer: _____

Where does she sing songs?
Answer: _____

What is Ava's favorite instrument?
Answer: _____

Word Search: Circus

H	N	Z	J	U	G	G	L	E	J	C
F	T	R	I	C	K	W	Q	S	Q	B
B	E	H	C	L	O	W	N	H	N	A
Z	N	T	A	H	Z	X	C	O	E	L
X	T	D	J	U	M	P	H	W	W	L
F	S	Z	S	Y	J	A	L	R	T	A
F	B	P	V	M	B	R	I	N	G	U
T	S	H	U	T	S	A	O	N	T	O
G	X	Y	U	V	V	D	N	E	Z	C
J	B	A	L	Y	P	E	U	W	L	U
Q	P	W	D	H	F	K	O	Z	C	L

Words:

Tent	Clown	Ring
Trick	Lion	Jump
Ball	Juggle	Show
Parade		

Rhyme Matching: Circus

Clown	Bent
Tent	Fact
Ring	Cat
Trick	Iron
Ball	Win
Net	Shade
Juggle	Tragic
Horse	Down
Acrobat	Pet
Show	Muggle
Jump	Giraffe
Magic	Sing
Tightrope	Course
Lion	Hope
Parade	Bow
Perform	Stick
Laugh	Call
Dance	Lump
Act	Warm
Spin	Chance

Rhyme Finishers: Circus

Lions and tigers, jump through **rings**,
The audience claps, and joyfully ____.

The ringmaster announces, with a voice so **grand**,
The greatest show, in all the ____.

Elephants parade, with such **grace**,
The circus, a magical ____.

Fire breathers, light up the **night**,
Circus acts, full of ____.

The circus tent, stripes red and **white**,
Where dreams come alive, every ____.

The trapeze artists, fly so **high**,
Daring and bold, they touch the ____.

Words:
Delight, Place, Sings, Sky, Night, Land

Spot 5 differences: Circus

Comprehension: Circus

Ethan went to the circus for the first time. Bright lights shone on acrobats swinging high. Clowns with big shoes made everyone laugh. A magician pulled a rabbit from his hat. And in the center ring, a juggler tossed pins in the air. The circus was a magical place for Ethan.

Questions:

Where did Ethan go for the first time?
Answer: _____

Who were swinging high under the bright lights?
Answer: _____

What did the clowns wear on their feet?
Answer: _____

What did the magician pull from his hat?
Answer: _____

What did the juggler toss in the air?
Answer: _____

Solutions

Solutions: Word Search

Animals

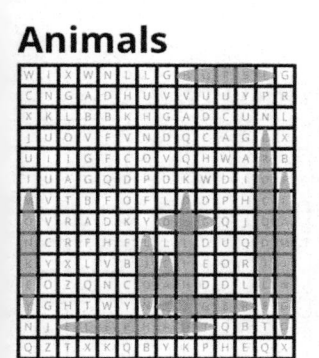

Space

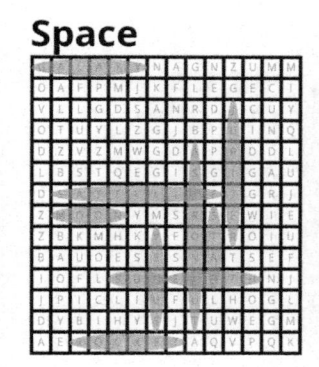

Nature

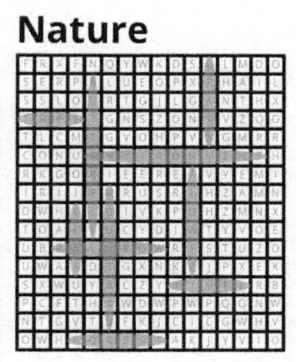

Food

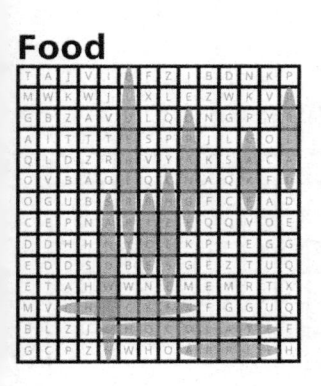

Sports

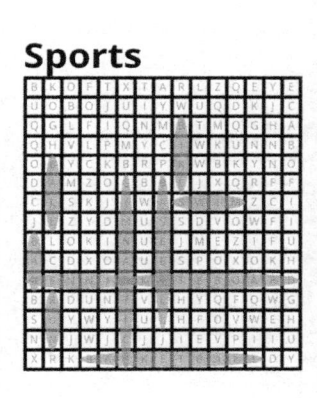

Transport

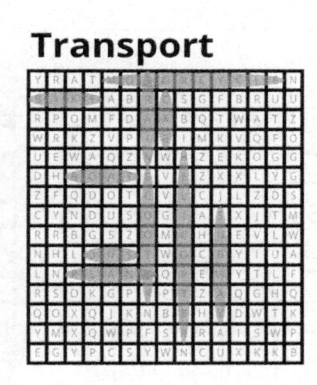

School

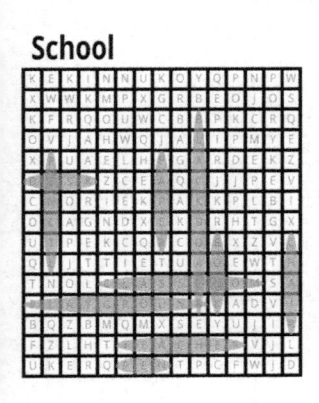

Weather

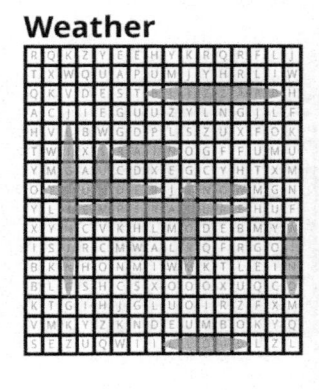

Holidays

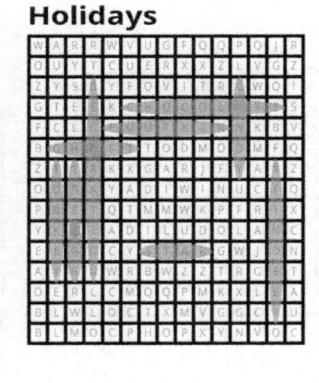

Solutions: Word Search

Fantasy
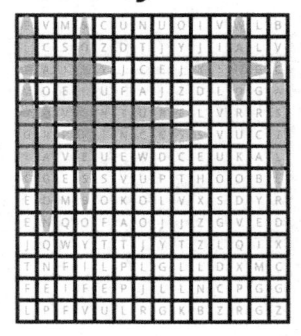

Dinosaurs
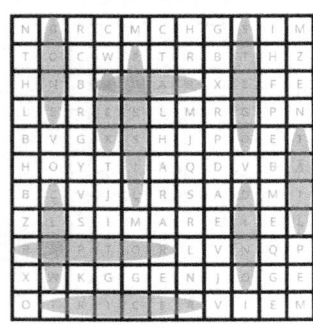

Under the sea
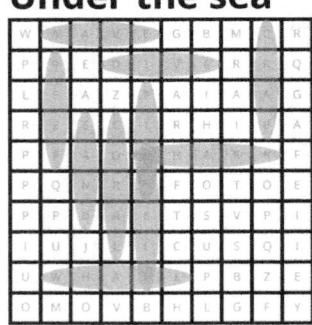

Superheroes
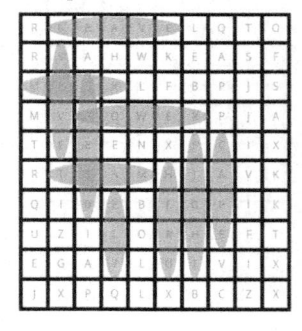

Farm Life
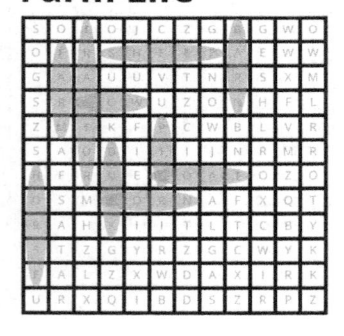

Pirates
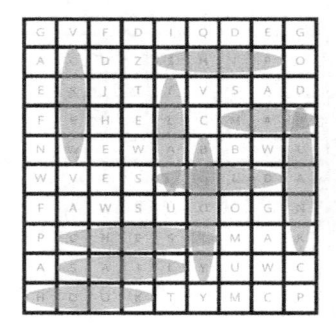

Construction
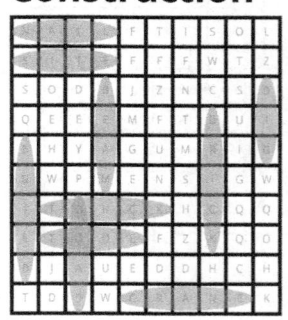

Robots
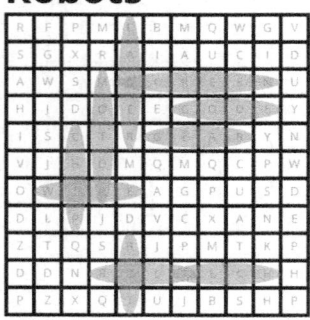

Jungle
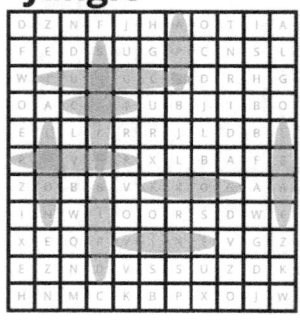

Solutions: Word Search

Music

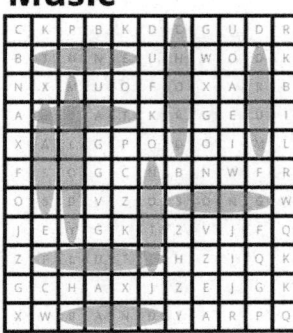

Circus

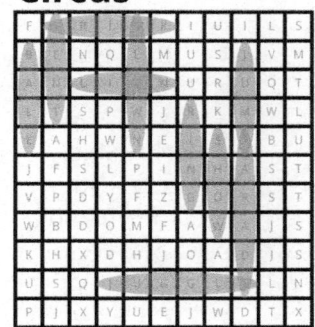

Solutions: Rhyme Matching

Animals	Space	Nature
Cat - Bat	Moon - Spoon	Tree - Bee
Horse - Course	Stars - Cars	Flower - Power
Bear - Hare	Sun - Fun	Leaf - Beef
Seal - Feel	Space - Face	Rock - Sock
Dog - Frog	Comet - Bonnet	Stream - Dream
Mouse - House	Mars - Bars	Sand - Band
Goat - Boat	Sky - Fly	Stone - Cone
Crow - Show	Astro - Metro	Cloud - Proud
Shark - Park	Rocket - Pocket	Mountain - Fountain
Lion - Iron	Galaxy - Taxi	Hill - Fill
Mule - Pool	Nebula - Scuba	River - Liver
Fox - Box	Orbit - Permit	Branch - Ranch
Duck - Truck	Star - Far	Bush - Push
Bee - Tree	Meteor - Detour	Grass - Pass
Swan - Lawn	Blackhole - Mole	Wave - Cave
Owl - Towel	Saturn - Pattern	Rain - Pain
Whale - Mail	Venus - Genius	Lake - Cake
Deer - Peer	Earth - Hearth	Wind - Kind
Snail - Pale	Eclipse - Clips	Snow - Crow
Gull - Bull	Planet - Granite	Sun - Bun

Solutions: Rhyme Matching

Food

Pie - Sky
Bread - Red
Rice - Nice
Meat - Beat
Soup - Loop
Cheese - Please
Bean - Green
Cake - Lake
Fish - Dish
Fry - Sky
Juice - Moose
Steak - Flake
Grape - Ape
Lemon - Demon
Berry - Cherry
Tart - Cart
Noodle - Doodle
Milk - Silk
Peach - Beach
Plum - Drum

Sports

Ball - Hall
Bat - Hat
Track - Back
Race - Base
Field - Shield
Score - Door
Swim - Limb
Goal - Hole
Kick - Stick
Dive - Hive
Tackle - Rattle
Run - Sun
Jump - Lump
Hit - Fit
Pitch - Ditch
Sport - Fort
Skating - Dating
Hurdle - Turtle
Dart - Cart
Row - Bow

Transport

Car - Star
Train - Rain
Boat - Coat
Plane - Crane
Truck - Duck
Ride - Slide
Fly - Sky
Bike - Hike
Van - Pan
Bus - Plus
Ship - Flip
Road - Toad
Rail - Snail
Drive - Dive
Taxi - Maxi
Metro - Retro
Yacht - Taught
Cycle - Tricycle
Scooter - Computer
Jeep - Peep

School

Book - Look
Pen - Hen
Chair - Bear
Desk - Nest
Note - Boat
Test - Best
Class - Glass
Math - Path
Learn - Fern
Read - Seed
Pencil - Stencil
Quiz - Whizz
Teacher - Preacher
Bell - Well
Board - Ford
Grade - Blade
Paper - Taper
Student - Prudent
Eraser - Chaser
Lesson - Cresson

Weather

Rain - Train
Snow - Bow
Cloud - Loud
Wind - Pinned
Storm - Warm
Fog - Dog
Mist - List
Sun - Fun
Thunder - Under
Lightning - Frightening
Drizzle - Fizzle
Chill - Hill
Breeze - Trees
Hail - Pale
Frost - Lost
Tornado - Avocado
Heat - Beat
Shower - Flower
Blizzard - Lizard
Dew - Shoe

Holidays

Gift - Lift
Tree - Bee
Treat - Sweet
Feast - Least
Card - Yard
Snow - Bow
Bell - Well
Santa - Atlanta
Elf - Shelf
Candy - Dandy
Easter - Feaster
Pumpkin - Dunkin
Ghost - Toast
Witch - Ditch
Turkey - Murky
Parade - Made
Firework - Quirk
Love - Dove
Costume - Plume
Bunny - Sunny

Solutions: Rhyme Matching

Fantasy

Knight - Light
Dragon - Wagon
Spell - Bell
Fairy - Berry
Castle - Tassel
Giant - Defiant
Wand - Pond
Mermaid - Parade
Unicorn - Torn
Gnome - Foam
Potion - Ocean
King - Ring
Queen - Green
Troll - Roll
Dwarf - Scarf
Elf - Shelf
Ogre - Poker
Jewel - Cruel
Magic - Tragic
Crown - Clown

Dinosaurs

Rex - Flex
Bone - Cone
Raptor - Actor
Tail - Pale
Claw - Straw
Stego - Lego
Roar - Door
Spike - Hike
Foot - Put
Scale - Snail
Dino - Vino
Beak - Peek
Wing - Sing
Fly - Sky
Tusk - Dusk
Hunt -Stunt
Teeth - Beneath
Wild - Child
Long - Song
Nest - Test

Under the Sea

Fish - Dish
Sand - Hand
Coral - Moral
Whale - Pale
Shell - Bell
Wave - Brave
Crab - Slab
Dive - Hive
Reef - Thief
Shark - Park
Squid - Lid
Swim - Limb
Boat - Goat
Fin - Tin
Tide - Hide
Star - Far
Deep - Coat
Float - Note
Blue - Shoe
Ray - Play

Superheroes

Cape - Tape
Mask - Flask
Fly - Sky
Hero - Zero
Power - Tower
Speed - Reed
Save - Brave
Team - Dream
Brave - Cave
Fight - Light
Wing - Ring
Quest - Best
Night - Right
Flash - Dash
Web - Ebb
Fire - Wire
Quick - Stick
Steel - Feel
Dark - Park
Vault - Salt

Farm Life

Cow - Plow
Hay - Day
Pig - Twig
Barn - Yarn
Duck - Truck
Corn - Horn
Goat - Boat
Horse - Course
Sheep - Leap
Chick - Stick
Mud - Bud
Farm - Arm
Hen - Pen
Stable - Table
Calf - Laugh
Field - Shield
Foal - Bowl
Tractor - Actor
Grain - Rain
Feed - Seed

Pirates

Ship - Slip
Gold - Bold
Chest - Best
Booty - Cutie
Plank - Rank
Flag - Bag
Map - Cap
Pirate - Private
Sword - Board
Hook - Look
Sail - Pale
Anchor - Lanker
Isle - Pile
Loot - Boot
Deck - Peck
Mate - Crate
Parrot - Carrot
Skull - Full
Ocean - Potion
Spy - Fly

Solutions: Rhyme Matching

Construction Site

Brick - Stick
Crane - Lane
Dig - Rig
Tool - School
Cement - Rent
Beam - Dream
Nail - Tail
Hammer - Stammer
Truck - Luck
Build - Filled
Dirt - Shirt
Block - Clock
Drill - Frill
Weld - Held
Bolt - Jolt
Lift - Drift
Pipe - Wipe
Road - Load
Work - Lurk
Crew - Blue

Robots

Wire - Fire
Chip - Slip
Bot - Cot
Code - Load
Screen - Green
Gear - Ear
Tech - Deck
Beam - Dream
Light - Fight
Port - Fort
Byte - Sight
Disk - Risk
Plug - Hug
Drive - Hive
Bolt - Jolt
Spin - Thin
Flash - Dash
Pulse - False
Track - Back
Grid - Lid

Jungle

Lion - Ryan
Vine - Pine
Monkey - Donkey
Tiger - Striker
Leaf - Beef
Snake - Lake
Jungle - Bungle
Bird - Word
Tribe - Vibe
Drum - Gum
Plant - Chant
River - Quiver
Path - Bath
Rain - Cane
Hut - Nut
Frog - Log
Swing - Ring
Roar - Oar
Claw - Straw
Wild - Child

Music

Note - Coat
Drum - Plum
Beat - Seat
Song - Long
Play - Day
Flute - Boot
String - Ring
Chord - Board
Band - Sand
Tune - Moon
Sing - Ring
Bass - Has
Loud - Cloud
Pitch - Ditch
Rock - Sock
Jazz - Grass
Blues - Shoes
Pop - Hop
Rap - Cap
Melody - Remedy

Circus

Clown - Bent
Tent - Fact
Ring - Cat
Trick - Iron
Ball - Win
Net - Shade
Juggle - Tragic
Horse - Down
Acrobat - Pet
Show - Muggle
Jump - Giraffe
Magic - Sing
Tightrope - Course
Lion - Hope
Parade - Bow
Perform - Stick
Laugh - Call
Dance - Lump
Act - Warm
Spin - Chance

Solutions: Rhyme Finishers

Animals

-Snore
-High
-Strong
-Tail
-Free
-Seat

Nature

-Day
-Around
-Hold
-Land
-Coat
-Crash

Sports

-Fast
-Ground
-Cool
-Beat
-Nice
-Feet

School

-Seat
-Side
-Me
-Fun
-Tell
-Small

Holidays

-See
-Crowd
-One
-Grow
-Year
-Night

Dinosaur

-Afraid
-Treat
-Sky
-Found
-Pride
-Be

Space

-Night
-Sight
-Spun
-Shine
-Great
-Sleep

Food

-Sweet
-Spice
-Eat
-Ice
-Heat
-Bite

Transport

-High
-Ease
-Fly
-Rain
-Be
-Everywhere

Weather

-Around
-Near
-Play
-Bright
-By
-Night

Fantasy

-Shriek
-Everywhere
-Air
-Be
-Past
-High

Under the Sea

-Way
-Night
-Sky
-Sway
-Around
-Seek

Solutions: Rhyme Finishers

Superheroes
-Fly
-Spare
-You
-Best
-Ease
-Go

Pirates
-Land
-New
-Skies
-Pride
-Fight
-Sun

Robots
-Fear
-Year
-Planned
-You
-Past
-Sound

Music
-Dear
-Move
-Cheer
-Boat
-Space
-Breeze

Farm Life
-Day
-Me
-Now
-Lot
-Day
-Sun

Construction
-Side
-Planned
-Place
-High
-Right
-Street

Jungle
-Free
-Sleep
-Line
-Told
-Stand
-Be

Circus
-Sings
-Land
-Place
-Delight
-Night
-Sky

Solutions: Spot the difference

Animals

Space

Nature

Foods

Solutions: Spot the difference

Sports

Transport

Schools

Weather

Solutions: Spot the difference

Holidays

Fantasy

Dinosaur

Under the sea

Solutions: Spot the difference

Superheroes

Farm life

Pirates

Construction

Solutions: Spot the difference

Robots

Jungle

Music

Circus

Solutions: Comprehension

Animals

-Zoo
-Eating leaves
-Monkey
-Splash water with its trunk
-Parrot

Nature

-In the forest
-Birds singing
-Rustled the leaves
-Jump into the water
-Flowers

Sports

-Jake
-Finished 2md
-Kicked a goal
-A high bar
-A blue ribbon

School

-Green Valley School
-Yellow
-Books, pencils, lunchbox
-Art
-Mia

Holidays

-Sophie
-Twinkling lights
-Gingerbread cookies
-In her living room
-Presents wrapped in colorful paper.

Space

-At night
-Stars
-Round
-Made a wish
-Astronaut

Food

-Fruit market
-Apples and bananas
-Red
-A tasty fruit salad
-Yes

Transport

-Vehicles
-Big buses
-Bicycles
-Train whistle
-Blue airplane

Weather

-Looked out the window
-Shining brightly
-Sway
-Puffy clouds
-Rain for the plants

Fantasy

-Land of magic
-A rock
-Dragons
-Animals
-Enchanted forest

Solutions: Comprehension

Dinosaurs

-Dinosaurs
-Sharp teeth
-Plants
-Velociraptor
-Dinosaur posters

Superheroes

-A superhero comic book
-Fly and shoot lightning bolts
-Flashgirl
-Faster than the wind
-Having superpowers

Pirates

-An old map
-An "X"
-A pirate hat
-A ship
-His toy parrot

Robots

-A toy robot
-Dance, light up, speak simple words
-Bright blue
-Rizo spins around
-Charging dock

Music

-Listen to music
-Ukulele
-Taps her feet
-At school
-Piano

Under the sea

-Snorkelling
-Colorful fish
-Jellyfish
-Dolphin
-Like she was in another world

Farm life

-On a farm
-Red
-Feeds chickens and collects eggs
-Plays with his dog, Spot
-Yes

Construction Site

-The construction site
-Dirt
-Heavy beams
-Helmets
-A builder

Jungle

-Jungle safari
-Tall trees
-Roar of a lion, chatter of monkeys
-Elephant
-The jungle was full of life.

Circus

-The circus
-Acrobats
-Big shoes
-A rabbit
-Pins

Printed in Great Britain
by Amazon

43467446R00066